DEDICATION

The work is dedicated to those of my family who kindled my eyesight and sowed the seeds of dream in my heart. They are my respected parents Late Sunil Kumar Mitra and Late Arati Rani Mitra and my dear grandfather Late Naresh Ch. Mitra.

Also I acknowledge my indebtedness to the honourable and respectable Prof. Dr. Pradip Kumar Mazumder, a great personality and a resolute artist of Mathematics who facilitated this repulsed author to publish this theory (Of mathematics) in their journal with joint endeavour of Dacca & Jahangirnagar University of Bangladesh. I pay my respectful gratitude to all of them.

Bongaon,   **– Manotosh Kumar Mitra**
24 Pgs (N), W.B.   Mobile: 9734695094

GRAMMAR OF MATHEMATICS & LINGUAL AFFINITY

THEMATIC GRAMMATICAL INTERPRETATION PRESENTING VOWELS, CONSONANTS & ALPHABET OF MATHEMATICS

MANOTOSH KUMAR MITRA

INDIA • SINGAPORE • MALAYSIA

Notion Press

Old No. 38, New No. 6
McNichols Road, Chetpet
Chennai - 600 031

First Published by Notion Press 2020
Copyright © Manotosh Kumar Mitra 2020
All Rights Reserved.

ISBN 978-1-64805-721-2

This book has been published with all efforts taken to make the material error-free after the consent of the author. However, the author and the publisher do not assume and hereby disclaim any liability to any party for any loss, damage, or disruption caused by errors or omissions, whether such errors or omissions result from negligence, accident, or any other cause.

While every effort has been made to avoid any mistake or omission, this publication is being sold on the condition and understanding that neither the author nor the publishers or printers would be liable in any manner to any person by reason of any mistake or omission in this publication or for any action taken or omitted to be taken or advice rendered or accepted on the basis of this work. For any defect in printing or binding the publishers will be liable only to replace the defective copy by another copy of this work then available.

CONTENTS

Foreword *7*
Preface *11*
Acknowledgements *13*
Prologue *15*
About the Author *17*

Chapter – 1   Necessity of Learning Grammar
(ব্যাকরণ জানার প্রয়োজনীয়তা) 19

Chapter – 2   Mathematics as a Language (ভাষা হিসাবে গণিত) 21

Chapter – 3   Sound and Letters – Vowels and Consonants
(ধ্বনি ও বর্ণ – স্বরবর্ণ ও ব্যাঞ্জণবর্ণ) 27

Chapter – 4   Philology/Phonetics/Science of Sound
(শব্দ-বিজ্ঞান) 31

Chapter – 5   Diphthongs (যুক্তাক্ষর) 34

Chapter – 6   Difference between Words and Parts of Speech
(শব্দ ও পদের পার্থক্য) 37

Chapter – 7   Particular Use of Words and Parts of Speech
(শব্দ ও পদের বিশিষ্টার্থে প্রয়োগ) 40

Chapter – 8	Case – Ending (বিভক্তি)	42
Chapter – 9	Compounds (Samasa) (সমাস)	45
Chapter – 10	Joining (সন্ধি)	54
Chapter – 11	Sentence, Subject & Predicate of a Sentence Arranging Parts of Speech in a Sentences (বাক্য, বাক্যে উদ্দেশ্য-বিধেয় ও বাক্যে পদস্থাপন)	56
Chapter – 12	Voice Change (বাচ্য পরিবর্তন)	60
Chapter – 13	Tense (কাল)	64
Chapter – 14	Punctuation Marks in a Sentences (বাক্যে অভ্যন্তরীন যতিচিহ্ন)	66
Chapter – 15	Substitution of Single Word or, One-Word Expression (এক কথায় প্রকাশ)	68
Chapter – 16	Indeclinables (অব্যয়)	71
Chapter – 17	Adjectives, Adverbs and their Classifications (বিশেষণ পদ ও প্রকার ভেদ)	72
Chapter – 18	Amplification (Of Ideas) (ভাবসম্প্রসারন)	76

Epilogue *83*

References *85*

FOREWORD

I am a Bengali and Bengali is my mother tongue. But my working/ professional life is settled as a teacher of Mathematics. My earnest effort comes from my love for the subject on one hand, and on the other, my sincere attraction to my mother tongue, Bengali, it's a bond between the two. So responsibility lies there. More to say from my long experience of teaching life that two fears are active in rural Bengal – during day it is the fear of mathematics and at night that of ghosts. Therefore, there remained a responsibility to extricate the students from the fear of mathematics. But how? Teachers are trying their best in their own capacity, still mathematics as a subject could not have acquired a favourite place in the minds of the students. A scary atmosphere still pervades as before. My journey starts just from this point.

Further I've noticed – the mother tongue of the Bengalis is Bengali, although its complex grammar has not created any fear in the minds of the students. The same argument can be placed in case of English language. Throughout the year they hardly provide time for Bengali or English, yet they score good marks in the subject by dint of trifle preparation, because of their mother- tongue. They must have natural command over that language. To acquire the art of rhetoric (ornamentation) takes some time. It can be observed, provided we keep our eyes and ears open, that learned people say 'Mathematics is a language'. But they fail to show cause in favour of the proposition: 'Mathematics is a language'. The best they can do to clarify mathematics as a language is to do by some mathematical equation or creating some problem of our real life.

My journey starts just from here and continues to the end where I've mentioned the mathematical equation of life-bound problems. I've therefore, tried to invest mathematics with the vision in the same way as Bengali language. It is to experience whether mathematics can be brought out as an allied force in the similar manner as grammar is an allied force to Bengali language; that is to if mathematics is also fully potent as grammar to Bengali language. With this, two purposes will be served by one method. It will prove that mathematics is a grammar potent language, and mathematics can be attributed as a Bengali-icon putting it in Bengali frame-work, so that fear for mathematics may disappear from the children or math-learner.

As for example – Bengali grammar contains letters, vowels and consonants, construction of words, stem, roots, parts of speech with case-endings, formation of sentences, voice-change, joining, compound(combined words), one word substitution etc, all are discussed in detail and explained to find Bengali as a dialect emerging from the angle of arguments. Basing on arguments Bengali has become powerful.

Exactly like that, I have tried to place mathematics as a language and to find out if it can be introduced chapter-wise vis-a-vis Bengali grammar as mentioned above. In fact, if a teacher imparts a comparative lesson on Bengali and mathematics simultaneously, then Bengali grammar will get the touch of mathematics, and mathematics, on the other hand, will get a room in the lap of mother-tongue, that is to say mathematics would be watched and explained through the window of Bengali grammar.

Suppose a teacher, during his discourse says, "Look, our mother-tongue is Bengali based on 52 letters, of which 12 are vowels and the rest are consonants. The vowels are a (অ), ā (আ).....to .... O (ও), au (ঔ) and the consonants are k (ক), kh (খ), g (গ) ....to ń (ং) as well as five vowels are in the English alphabet and rest twenty one are consonants." While math-language is based on ten letters only of which there are only two vowels 0 & 1 and there are eight consonants: 2, 3, 4, to 9.

In mathematics 'letter' is taken for 'digit' and 'word' for 'number'. The next to be done is to clarify the technique of word formation, differentiate word and parts of speech, and that the success of mathematics is hundred percent more than that of Bengali in word – formation. And then, if joining of words is placed vis-à-vis the compounded form of number, mathematics will easily be understood and the fear will diminish mostly. The above matters are fully discussed in chapters later on.

The rules of mathematical languages are in its own place. It is only within a certain boundary of mathematically experienced intellectuals and academics, where mathematics is being explored, explained and exploded and only from mathematical view point this is not subject to constraints of other languages.

Therefore, I have tried to increase the magnitude of this limited boundary as an easy way to establish mathematics as a language. It is from this belief, faith and responsibility that the touch of mother-tongue should become a bit more vital to learn mathematics fearlessly. That is by showing mathematics as a clone of Bengali language; perhaps the fear of mathematics will be overcome. Since Bengali & English languages are almost identical, this attempt is also applicable to English language. And this concept applies to all the languages of the world.

PREFACE

The book, 'Grammar of Mathematics and Lingual Affinity' by Manotosh Kumar Mitra, the retired Headmaster and researcher of mathematics is a research–guide, embellished with fundamental thinking. As a subject it is not merely to highlight a creative effort under modern view. The math- devoted author has presented the inextricable miscellaneous applications of mathematics through trifle skilful experience by using the language of day-to-day conversation and the experience of our pragmatic life. He has upheld the domain of mathematics deeming with the elegance splendid subject- matter in the radius of perfect affinity with Bengali language. He has termed the two letters 0 & 1 as vowels, and 2, 3,............ to 9 as consonants . The term is perfect.

Bengali is our vernacular and mother-tongue. Basing on the grammar of this language he has arranged chronologically the field of his thinking. A glance at the chapters reveals how sincerely he has strung his thinking at the root of the subject-matter. The respected author has at first forwarded his own endeavour is composing this book to meet the basic needs. He has arranged the basic matters following the pattern of school level Bengali grammar – sound and letter, philology, case endings, compound (combined words), joining, voice-change etc. The treatise has nowhere been over-loaded. The author has presented the subject-matter briefly in an easy, lucid and realistic way.

An inquisitive reader may or may not concede to the discourse of a subject. In both the cases there remains, of course, the scope of further discussion and deliberation. Experts in mathematics and language need to sit abreast to promote the subject to the international arena. The author's efforts will attain success if it is materialized. This optimism is quite pertinent.

The "PASCHIMBANGA GANIT PARISHAD" has intently published the book in Bengali with a view to expand the subject orientation. The more free, independent and fundamental thoughts are brought out and spread out, the more the waves of thinking on the subject are diffused to draw the attention of radical thinkers. The scope of cognition regarding the science of mathematics in Bengali is still limited. All our efforts are dedicated to root firmly the manifestation of thoughts on mathematics.

Hope the thought on the subject will spring up as the source of intellectual device through constructive deliberation.

Kolkata

– **Dr. Pradip Kumar Mazumder**
PASCHIMBANGA GANIT PARISHAD

ACKNOWLEDGEMENTS

I would like to express my gratitude to the honourable GOUR BINOD DAS, respected retired teacher, for translating my book into English. I sincerely acknowledge his contribution.

PROLOGUE

During the intervening period of responsible work of teaching I have seeded the words of those who have been dwelling in the shape of God spanning in my heart forever. They are my very affectionate and beloved students. Sometimes, I have talked a bit more playfully keeping myself associated with them. Time stored up those spent words carefully. Those (words) were nurtured in the time's womb and are on the way to be released in a more embellished form. I offer this new-born to the hands of god incarnate readers to take care of.

In 2000 A.D the theory was first published in a concise form in the preface of 'Ganit Parikrama' by the joint patronage of University of Dhaka and Jahangirnagar University with the financial help of Bangladesh Govt. Then in 2002 it was published in the preface of Ninth State Science & Technology Congress arranged by Viswabharati University as a research-synopsis. In the same year it was published in a miniature form in the Magazine, Jnan-Vignan conducted by Bangiya Bijnan Parisad.

I tender my gratitude to Mr. Dipak Kumar Daw, the respectable Headmaster and an extra-ordinary entrepreneur, without whose liberal help the effort could not the materialized. I admit my indebtedness to respectable Jaydev Dey, an indefatigable science worker, as I have got the opportunity to get back in the field of mathematics with their all-wide assistance and exchange my experience and knowledge with the readers.

I shall consider my efforts successful provided the object of the book achieves this goal.

Bongaon,   – **Manotosh Kumar Mitra**
West Bengal   Mobile No: 9734695094

ABOUT THE AUTHOR

Manotosh Kumar Mitra, retired headmaster was born at Hadipur, 24 Parganas (North), West Bengal. He engages himself in new fundamental works on Mathematics. His first utterance of the terms 'alphabet', 'vowels', 'consonants' in mathematical language may open a new window to Mathematical culture. The synopsis of this content was first published in their journal of University of Dhaka and Jahangirnagar University in Bangladesh in the year of 2000, and then in State and Science & Technology Congress in 2002. He has been researching on pedagogic analysis for the last 25 years with untiring efforts to visualize the abstract facets of mathematics. He is currently conducting mathematical workshops on mathematics in the school level to remove fear from the minds of the students. This book is already published in Bengali version by 'PASCHIMBANGA GANIT PARISHAD'. Beyond this a good number of books by the author have already been published.

Front Matter

Mathematical alphabet, vowels & consonants are the first grammatical interpretations and analyses on the subject and are of the first pronunciation in the world.

Chapter – 1

NECESSITY OF LEARNING GRAMMAR
(ব্যাকরণ জানার প্রয়োজনীয়তা)

If anyone wants to learn Bengali, English or Mathematics, he must know the grammar. For the sake of language he should learn the letters and alphabet used in language, number of letters in the alphabet, difference and relation between the vowels and the consonants, technique of word formation, how the words become parts of speech added with case ending, and how the parts of speech being placed side by side make a sentence and express our minds.

I do not like to entertain this absurd question-which one is more important between learning grammar for the sake of language and learning language for the sake of grammar…? It is healthier to approach by thinking the one as repletion of the other. The foundation of our mother tongue, i.e. Bengali is grammatically potential, and to assimilate the language correctly there are sufficient potent grammar books. So let us see – what grammar is. Men used to live a wild life at the early stage of creation of the society. In that early stage they used to express their mental feeling through gestures and postures for their livelihood. Exchange of mental feeling gave birth to language – through much evolution. Language got a shape of flowing stream after content use by human mouths like the waterfall. Gradually, being bathed in the womb of time it got a refined, scientific, universal, read and written form. In this way in respect of time and place different language came to serve the need.

Moreover, scientific grammar with rules and principles, became necessary for the pure assimilation of language, learning which one can help teaching not only one's own community, but alien people also. That is to say, the book in which rules and principles are scientifically and integratedly dealt with, to speak, write, understand and make understand the language is called grammar.

In the similar background, a beautiful, scientific and grammar based language is the object of my search and need for writing. My initiative aims at delving into Bengali grammar and finding if it is wholly applicable to mathematics.

Actually, we have so far looked upon mathematics as a subject confined in a classroom, and dealt with it in the same scale as other subjects. Consequently, the application method of mathematics is so artificially manifested, and presentation is so faulty that in many respects the subject fails to root itself as a grammar. My effort of writing is, therefore, to bring out mathematics in cogency with language.

Chapter – 2

MATHEMATICS AS A LANGUAGE
(ভাষা হিসাবে গণিত)

Language is of unlimited importance to provide good health to the members of a civil society. Bengali language is created for the exchange of feeling of the Bengalis. English language is created for the exchange of feeling of the English. Alongside this 'Math–Language' was created for the exchange of a different feeling in the minds of not only the Bengalis or the English, but all the races. It is not confined to mere provincialism. Its emergence is universal and worldwide, and used by the multitude. Without learning it correctly no race can climb the summit of prosperity – it is undeniable. Even in the daily life of man the beautiful use of math – language is noticed.

For instance–dinner at night and defecation in the morning – are a daily necessity of human life. It is as, it were, a nice reflection of mathematical application in each life. Notwithstanding the instance being coarse with the touch of humour, the matter is perceived completely and successfully.

i.e. Life + food = lunch/dinner

life – indigestible part = faeces

The above two life – equation (phrase) exist as an everyday mathematical touch in each life for livelihood. Each step of movement for going from one place to another place, proceeds following mathematical formula, because the interim space between the two legs is a constant quantity. Particular mathematical formula is revealed if each step is marked. It silently gives birth to a particular chapter of mathematics called arithmetical progression.

So to string the practical life with one thread a new language has come of 'mathematics'. Otherwise to say– 'life is mathematicious'.

Besides the language we always interact in, is touched with mathematics unnoticed. The simultaneous fettering of mass-ability symbolism and overall composition brought forth a different language, called mathematical language: e.g. - Once the students said, "Sir, more than half the students have left." Its mathematical form is: Numbers of students left $y > x/2$, where number of students present were $= x$. From such statement math- language became more different and capable of speaking multilingual words: e.g.: $y - x/2 > 0$ or $x/2 - y < 0$, or $2y - x > 0$, or $x - 2y < 0$.

Each of the above inequations brings out a complete sentence. Moreover, when we say 'I have passed a greater part of life' or, 'I understand the matter more than you'- these are some sentences touched with inequation. Its mathematical form is: $y > x/2$: $y =$ spent life & $x =$ entire life. The first sentence is expressed otherwise by some people: such as 'less than half of life still remains". Its mathematical form is: $x/2 - y < 0$. What else is there to say in Bengali or in English? Possibly nothing. But in mathematical language the above equation can be expressed in more different way: such as: $2y - x > 0$ or $x - 2y < 0$ or $x < 2y$. The second sentence above may similarly be expressed in many mathematical languages. Thus, innumerable examples may be given; such as: My present age is double the present age of yours $\Rightarrow x = 2y$, where x & y are the present age of mine and yours respectively.

My present age is more than double the age of yours $\Rightarrow x - 2y > 0$ where x & y are the present age of mine and yours respectively.

My present age is 5 years more than the double the present age of yours ⇒ x = 2y + 5, where x & y are the present age of mine and your respectively.

Every moment, the incidents in our life-stream, spoken words, or life-style, whatever we say are expressed mathematically. And when it is rightly expressed, it becomes a language or multipurpose presentation, a mute revolution in symbolic synopsis. Mathematics is, therefore a language. It is a symbolic language, although all languages are partially symbolic in reading and writing, such as ক (ka) read symbolically as ক (k) + অ (a) = a unit of two letter, ক (k) is included in consonants and অ (a) is included in vowels.

If mathematics is placed alongside this, then it is also a grammar – potent, expressive and brilliant language with letters, alphabet, vowels and consonants. As in Bengali language there are fifty (52) letters ..... from অ (a) to ̇(ṅ) , among which there are twelve (12) vowels: অ (a) to ঔ (au), and the rest are consonants amounting to 40. In English language there are five vowels such as a, e, i, o, u and rest twenty one are consonants. Keeping mathematics beside it the comparison reveals that the numbers of letters in math-language is ten (10) only. Out of these ten letters there are only two vowels and the rest eight (8) are consonants. The letters are: 0, 1, 2, 3, 4, 5, 6, 7, 8, 9. 1 to 9 are original or natural. But '0' is a newly created letter (digit), of which '0', '1' are vowels and 2 to 9 are consonants.

The writing symbol of two letters is '2' and reading symbol is two, the writing symbol of three letters is 3 and reading symbol is three etc. Sitting side by side these letters make words. As in Bengali ক (ka) + র (ra) = কর (kara) and সিংহ (sinha) + আসন (āsana) = সিংহাসন (singhasana), so in mathematics = 2 + 7 = 9 or 11 + 15 = 26 . The first is letter (digit), after being joined a new word (number) is made. And the second word after being joined a new word is made. Again these words being added with case – ending and sitting side by side give out a mental feeling by making a sentence (equation). Then the language comes of.

As in Bengali – 'তোমাকে দেখতে ঠিক গোপালের মতো' (you look exactly like Gopal). In this sentence the word তুমি (tumi/you) has become তোমাকে (tomake/to you) with কে (ke) ending, and the word দেখা (dekha/look) has become দেখতে (dekhte/to look) with ইতে (ite) ending; thus they have become appropriate parts of speech. That is তুমি (tumi/you) + কে (ke) = তোমাকে (tomake/to you), দেখ (dekh/look) + ইতে (ite) = দেখতে (dekhte/to look). These words being added with different ending have become appropriate parts of speech of a meaningful sentence.

Similarly in mathematics too, $20 + 35 = 55$ is a word-joining. Here 20, a number (word) and 35, a number (part of speech) being added make a meaningful word. 20 is a word; it is a stem (প্রাতিপাদিক) and with no ending. And $30 + 5 = 35$, i.e. a new word 30 (stem) being added with 5 (ending/বিভক্তি) makes a new part of speech, which is applicable in a sentence as well as equation.

Again, $2x + 3y = 75$ is a mathematical equation, which expresses a particular relation with two variables. Here the two variables x & y are two nominal. It, being multiplied by 2 & 3, has turned into an applicable sentence (equation) after being added with root (verbal) ending. Any mathematical equation is, therefore, actually a sentence. So it can be said summarily that mathematics is a language.

Keeping in front the expansion of all the alphabets used in different languages of the world, the expansion of their grammar and depth, it may be said that mathematics is the only science – based language. In view of the number of letters, the alphabet of math-language contains the lowest numbers of letters and its grammar is easy to grasp. The span of its use is so far-flung that the ancient scholars say about it: "Mathematics is the mother of all sciences." And advancing a few steps more we can say that mathematics is the mother of all languages existing in the universe. So in a word it may be said – 'Mathematics is a grammar-potent language.'

Again, as in the alphabet each letter has its particular meaning and position, in mathematical alphabet too the same rule is applicable as–

the vowel অ (a) means 'No' or makes anything negative, and its position is first in the alphabet ক (ka) means water and its position among the consonants is first. খ (Kha) means sky, and its position among the consonants is second etc. Likewise each letter (digit) in math alphabet is a successful representative form of countability. The meaning and position of each letter is specific. But other than Mathematics, in the alphabet of any language the position of letters depends generally on counting fingers, meaning is also set upon excessive cramming. But in the alphabet of math language the meaning and position of letters are very clear and dependent on instant comprehension. There is no need to grope about in the mathematical alphabet going down memory steps. Again in Bengali alphabet there is much complexity in vowels. In terms of time (taken to pronounce) some are short and some are long vowels: as short (হ্রস্য) ই (i) and উ (u) and long (দীর্ঘ) C (ৱ) and উ (ū). Alongside this, the vowels of mathematics are completely free from the above complexity. Among the consonants the first and the third letters of each unit (বর্গ) are unaspirated (অল্পপ্রাণ), and the second and the fourth letters are aspirated (মহাপ্রাণ). Side by side, in math-language odd positioned numbers (letters) are even, and even positioned numbers (letters) are odd. The mathematical alphabet are 0, 1, 2, 3, 4, 5, 6, 7, 8, 9. So it is easily seen that odd position numbers (letters) are 0, 2, 4, 6, ... etc. are even and even positioned numbers (letters) are 1, 3, 5, ... etc. are all odd. খ (Kha) is the aspirated form of ক (ka), and ঘ (gha) is the aspirated form of গ (ga): i.e. খ (kha) = ক (ka) + হ (ha); ঘ (gha) = গ (ga) + হ (ha). Exactly in the same way it is seen that even letter (numbers) = odd number (letter) + 1 + or, odd number (letter) = even number (letter) +1. For example $3 = 2 + 1$, $4 = 3 + 1$. Similarly, even letter (number) = odd number (letter) −1, or odd number (letter) = even number (letter) -1 . For example: $2 = 3 − 1$, $3 = 4 − 1$ etc. i.e. the difference of two consecutive even and odd letters is = 1 (one). Besides, in respect of consonants in Bengali language the first and the second letters of each unit (বর্গ) are surd (অঘোষ বর্ণ), and third, fourth and fifth are sonant (ঘোষ বর্ণ). Rightly alongside 2, 3, 5 & 7 of math consonants are prime numbers, the rest are artificial, the digit '6' is both artificial and perfect: $1 + 2 + 3 = 1.2.3 = 6$

Remember, in the alphabet of math-language each succeeding letter (number) is the aspirated (মহাপ্রাণ) form of the preceding letter: i.e. $6 = 5 + 1, 7 = 6 + 1$ etc.

Chapter – 3

SOUND AND LETTERS – VOWELS AND CONSONANTS
(ধ্বনি ও বর্ণ – স্বরবর্ণ ও ব্যাঞ্জনবর্ণ)

Scientific base/foundation: Sound is the chief constituent of any language. Language emerges from sound. Sound-combination expresses mental feeling. To give a written form of this feeling and language, script or alphabet was created.

In the earliest stage of creation men used to express their mental feeling through picture-script. Later, in course of time, this picture script gradually changed into sound-script or alphabet and assumed a particular form acknowledged and accepted by all, formally christened 'letter'. The assemblage of such useful letters formed the alphabet, that is: the written form of sound is letter. These letters are the foundation-stone of word. The assemblage of meaningful letters makes a word. Being added with endings the words become parts of speech. And when the parts of speech are arranged together and express mental feeling, and make the listeners react, then it is a sentence. Thus languages come of whether it is Bengali, English or any other language of the world. These are merely the repetition of familiar facts.

It is referred to with a view to establishing our objective that 'mathematics is a different language, and that too following the Bengali grammar in toto.'

Now let us see the scientific position of each in respect of the alphabet of Bengali and math-language and how much they are successful in expression. In Bengali alphabet there are 52 (fifty two) applicable letters, of which vowels are অ (a), আ (ā) to .... ও(o), ঔ (au) = 12 (twelve) in all, and consonants are ক (ka), খ (kha), গ (ga) to ....(ń) = 40 in all.

Similarly in English alphabet there are 26 (twenty six) applicable letters, of which vowels are 'a', 'e', 'i', 'o', 'u', five in all and consonants are b, c, d, f... etc. twenty one in all.

Alongside this in matter of math-language the letters (digits) used in the alphabet of mathematics are 0, 1, 2, 3, 4, 5, 6, 7, 8, 9 respectively = 10 in all, of which from 1,2, to 9 all are natural and '0' created to meet the need, is a new letter (digit), i.e. in the math-alphabet, the number of letters (digit) is ten in total from '0' to 9, of which '0' & '1' are vowels, and the rest i.e. from 2 to 9 = 8 (eight) in all are consonants. In modern algebra we see that zero (0) and (1) are the additive identity and multiplicative identity respectively. Just using these as vowels from its character point of view.

Now in view of number, the number of letters used in the alphabet of mathematics is far less than those used in Bengali, or English. It (number) is only ten, not over-burdened. From this perspective the success of mathematics is very scientific.

To learn the language having excessive number of letters is not only time absorbing, but unambiguously more intricate for teaching and hard for possessing by the brain. So it is proved that in view of the number of letters contained in the alphabet, Math-language is much more scientific.

Vowels and Consonants: From the angle of independent pronunciation of letters the alphabet is divided into two: From অ (a) to ঔ (au) are vowels, and from ক (ka) ......to (ń) are consonants.

In the alphabet the letters which are pronounced independently or deserving the power of own accent, have got the symbolic name 'vowels' as in Bengali অ (a), আ (ā) .... to ঔ (au) = 12 in all; of course, a few of them are now obsolete. And the letters which cannot be pronounced independently are called consonants, such as – ক (ka), খ (kha), গ (ga) to ....(ń), as they are pronounced: ক (ka) = ক (k) + অ(a); খ (kha) = খ (kha) + অ (a); i.e the letters ক (ka), খ (kha) etc. do not deserve the power of own accent. So they are consonants.

In math language, out of 10 letters (digits) only two are vowels since they have the power of own accent. They are '0' & '1", the rest 2 to 9: these eight letters are consonants. As the consonants cannot be pronounced without the help of vowels, so in math-language consonants: i.e. 2 to 9 cannot be pronounced without the help of vowels i.e. '0' & 1.

e.g.: ক (ka) = ক (k) + অ (a); গ (ga) = গ (g) + অ (a).

In mathematics $-2 = 2 + 0$ or 2×1

$3 = 3 + 0$ or 3×1 etc;

That is no letter (digit) can be pronounced or written without '0' or '1'. So '0' & '1' are vowels and the rest: 2 to 9 are consonants.

'0' is called the identity of addition and '1' is called the identity of multiplication.

In modern algebra, we see that zero (0) and one (1) are additive identity and multiplicative identity respectively. So zero and one can easily be considered as vowels and the rest from 2 to 9 are consonants.

The existence of letters and alphabet in math – language are exactly the same as in the first phase of grammar in any language. With the assemblage of vowels and consonants the math- alphabet is a mighty claimant to set up mathematics as a language.

Math-language is similar with the letters and alphabet, vowels and consonants of other languages of the world. And to follow the designated rules of math-grammar is a comprehensive principle.

Chapter – 4

PHILOLOGY/PHONETICS/ SCIENCE OF SOUND
(শব্দ-বিজ্ঞান)

Now let us come to judge the extent of lingual and scientific potentiality of Bengali and math-language, on the basis of formation of words constituted by letters used in the alphabet.

The number of letters in Bengali alphabet is fifty two (52). Let us take three of them – ক (ka), ম (ma) & ল (la). From the arranged of these three letters or, if we permute the three letters taken three at a time, six words may come off as কমল (kamal/lotus), কলম (kalam/pen), মলক (malak/meaningless), মকল (makal/meaningless), লকম (lakam/meaningless), লমক (lamak/meaningless). So two of the six words: i.e. কমল (kamal/lotus), কলম (kalam/pen) are meaningful and the rest four words মলক (malak), মকল (makal), লকম (lakam), লমক (lamak) are all meaningless. Hence the Bengali language (i.e. alphabet) burdened with 52 (fifty two) letters is not free form the weakness of word formation. Similar example can be cited taking three (3) letters from twenty six (26) letters of English alphabet constituted with letters half of Bengali alphabet. Let us consider 'c', 'a', 't' out of twenty six letters. Now if we permute the three taking three at a time we get six words e.g.; 'cat', 'act' – these two are meaningful and 'atc', 'cta', 'tca', 'tac'- these four are meaningless.

This is just a problem of the meaning of words constituted with three letters out of a big store. When the number of letters is four instead or three or five instead of four, there is the problem: i.e. the more is the number of letters, the more is the problem of meaning. This is a problem of comprehension in matters of the recognized languages like Bengali, English and others.

But along-side this, in math-language the ten-letter (digit) alphabet: 0 to 9 is completely free form the above weaknesses in creating word.

Number of letters used in math alphabet is ten only. Out of these ten, suppose any three are taken: 1, 2, 3. Out of these three digit, when arranged, or, if we permute the three letters taking three at a time, six meaningful words (numbers) ensue; none of them is meaningless: as 123, 132, 231, 213, 312, 321. Each word (number) bears different meanings. i.e. without being burdened with unnecessary letters. Math-language is very successful and scientific in creating words. Moreover, in mathematics words (numbers) the arrangement of letters, as many as one likes, will increase meaningful words at a higher rate, not a single meaningless word will in intrude.

Thus math-language is completely free from the above weakness of Bengali or English language in framing words. It is proved that math-language is the only scientific language in matters of producing meaningful words by using the letters of the alphabet.

To establish mathematics as a language we have got letters (digits), alphabet (numbers), vowels, consonants and words (numbers). We can briefly say that Mathematics has caused a silent and trouble free revolution in the world of languages to create words (numbers).

To constitute a language, words have the chief role. On the basis of word – creation scientific activities of the letters (digits) in the alphabet are not only wonderful, but also very reasonable and prefatory to successful presence.

So the first step to establish mathematics as a language is complete.

In case of number (words) having two letters (digits) the aggregate of multiplication and addition is maintained by a particular formula.

A single digit number being multiplied by one and a ten digit number being multiplied by ten become a double digit word (number) after addition: such as the number, 23. 23 come as an aggregate of 2 multiplied by 10 and added with 3: i.e. $23 = 2 \times 10 + 3 \times 1$; $35 = 3 \times 10 + 5 \times 1$. Again the same formula is applied in case of three digit number. Suppose in the number 327 the digit in the hundred place is 3, the digit in the ten place is 2 and the digit in the one place is 7. Consequently 3 multiplied by 100, 2 multiplied by 10 and 7 multiplied by 1 bring out the above number: i.e. $327 = 3 \times 100 + 2 \times 10 + 7 \times 1$.

With this sole formula four, five, six............ etc, all digital numbers can be created i.e. the word (number) will ensue as a result when, with the place value of each of the letters (digits) is added after multiplication. And each word is original in the scale of its meaning, and bears different meanings.

So far I have focused on the nominal character of words (number). Now let us come to judge the conjugation of words. In Bengali grammar কর্ (kar/do), চল্ (chal/go), বল্ (bal/speak), শুন্ (shun/listen) etc. are the verbal roots. Being added with numerous ending they become parts of speech, and are used in sentences: such as কর্ (kar) + ই (i) = করি (kari/do), কর্ (kar) + ইলাম (ilam) = করিলাম (karilam/have done), কর্ (kar)+ ইব (iba) = করিব (kariba/shall/will do). Here is original stem or root is কর্ (kar/to do). With that root are added ই (i), ইলাম (ilam), ইব (iba) etc. ending and make original (radical) meaningful parts of speech. Again in mathematical polynomial the example of uses of numbers; in such a manner such that $(2 \times 5) + 1 = 11 = 10 + 1$ or $(5 \times 6) + 2 = 32 = 30 + 2$. In each case there is the clear touch of verbal ending.

Now let us take a mathematical equation. Suppose the equation is $3x + 2y = 50$. It is a complete sentence (equation). $3x$ & $2y$ are parts of speech used in a sentence. These parts of speech are adjectives to nouns.

Chapter – 5

DIPHTHONGS (যুক্তাক্ষর)

In Bengali language we can see the application of diphthongs. Such as the word শব্দ (savda) contains diphthongs. Diphthongs are created to make word more meaningful to annex in writing and to shorten the word. Once Bengali language was very much influenced by diphthongs. Of late a new trend has come up to segregate or separate diphthongs and write the word with separate syllables with a view to simplifying and spreading mass literacy or education: e.g. তক্তা (Takta), তপ্ত (Tapta), যন্ত্র (Yantra) etc.

Now তক্তা (Takta) is written as তকতা (Takta)

তপ্ত (Tapta) is written as তপ্ত (Tapta)

যন্ত্র (Yantra) is written as যনতর (Yantra)

The new one comes of the habit of acceptance. So some people prefer diphthongs, and some again seek for the need to write segregating them.

However, in mathematical alphabet too diphthongs are widely used. In a sum for division, the divisor is less than dividend, and if the division is not exactly divisible, then the division process will said to be and improper fraction. And like the Bengali diphthong the number we get is similar to diphthong. Such as: if 15 is divided by 4, the quotient is 3 and remainder is 3 and the result is $3\frac{3}{4}$ which in mathematical language is called improper fraction (i.e. numerator greater than denominator). But keeping the Bengali word তপ্ত (Tapta) beside it, if compared, they seem quite alike.

Let us see তপ্ত (Tapta) $3\dfrac{3}{4}$

তন্ন (Tanna) $3\dfrac{2}{3}$

অভিশপ্ত (Abhisapta) $321\dfrac{3}{4}$

That is the application of diphthong is seen in math-language. But in modern Bengali diphthongs are written after segregation. Such as তপ্ত (Tapta) is being written তপ্ত (Tapta), তক্তা (Takta) is being written তক্তা (Takta) etc. Now if the question arises- can we likewise write the diphthong and improper fraction in math-language?

Then let us see: $3\dfrac{3}{4} = 3 + \dfrac{3}{4}$
$= 3 + 0.75$, or, $3\dfrac{3}{4} = 3.75$

We can see perfect affinity of segregated writing between Bengali and math-language. So Mathematics can briefly be called language. A question may be raised in this context – in the familiarity of expressing diphthong which language is the pioneer? i.e. Which on has copied the other? Now only that, in Bengali language there is no particular rule of using diphthongs. Use of diphthongs in words can be in any situation: Such as – in the word দ্বন্দ্ব (Dwandwa) diphthong is used both in the first and the second places, in the word তিক্ত (Tikta) diphthong is used in the second place, in the letter উত্তপ্ত (Uttapta) diphthong is used in the second and the third places, again in the word অভিশপ্ত (Abhisapta) diphthong is used in the fourth place etc. That is, in Bengali language, as there is no particular rule regarding the position of diphthong, teaching of diphthong is really complicated. (Bengali language) enjoy the advantage only because it is our mother – tongue and since we are the Bengali race. But to the foreigners the learning of diphthongs remain a bleak lesson, but vis-a-vis in math-language in matters of conjunct words (improper fraction) diphthongs remain in last position of the word (number) (in the tenth place after the one place), nowhere else: such as $3\frac{3}{4}$ or $123\frac{3}{4}$ etc, it can never be written $12\frac{3}{4}$ 32, i.e. in matters of mathematical conjunct words (numbers) a common formula in used, which is not only scientific, but it makes teaching – learning atmosphere very easy. Moreover it is noticed that there is no single lettered diphthong in Bengali language: such as ত্ব (Twa), ত্ন (Tna) or ক্ক (kka) are non-existent, but in math- language, alongside this, there is use of large number of one lettered diphthongs: such as 2/5, 2/7 etc, and these are independent and meaningful one lettered word. Not only that, the joining of two one lettered diphthongs is not permissible in Bengali grammar. But in Mathematics the joining of two one lettered diphthongs is quite easily worked out: such as $1/3 + 2/3 = 1$; $\left(\frac{2}{3} + \frac{3}{7}\right) = \frac{23}{21} =$. So, from grammatical point of view math-grammar is more scientific than Bengali grammar or English or other grammar.

Chapter – 6

DIFFERENCE BETWEEN WORDS AND PARTS OF SPEECH
(শব্দ ও পদের পার্থক্য)

The synthesis of meaningful sound that comes out of human voice makes the word, i.e. meaningful sound – combination is called word. In linguistics it is called stem or root. The word বসুমতী (vasumati) is constituted with the combination of the sounds ব্ (v) ,অ (a), স্ (s), উ (u), ম্ (m), অ (a), ত্‌(t) ঈ (ī) . The word has many meaning as পৃথিবী (prithivi/earth). ধরণী (dharani/earth), মেদিনী (medini/earth), অবনী, (avani/earth), ভূ (vhu/earth), ভূমি (bhumi/earth), বসুন্ধরা (vasundhara/earth) etc. Again, if the sounds 'ব', 'স্', 'ম', 'তী' change their places, they suffer from meaningless. In Bengali language it is a distinctive weakness.

One syllabic word কে (ke), ও (o), সে (se), বা! (bah!)...... etc. multi syllabic words -রতন (Ratan), কাদের (kader), আমাদের (Amader) etc. Words without ending are called প্রাতিপাদিক or প্রকৃতি (pratipadik/or prakriti, i.e stem or root) বাড়ি (badi/house or home), ঘর (ghar/room), গাছ (gach/tree) etc, are নাম-প্রকৃতি (Nam prakriti/stem) , and চল্ (chal/go), শূন্ (sun/listen), বল্ (val/speak) etc. are verbal roots.

Earlier we have come to know that in math-language there are ten meaningful sounds or letters (digits). These are letters (digits) as well as one lettered word and one digit number. The letters are 0 to 9; and after 9, 10 to 99 – all the numbers (words) are double/two digit numbers like two lettered words. These are (99–10) + 1 = 90 in all. After this from 100 to 999 are three digit numbers like three lettered words.

These are (999 − 100) + 1 = 900 in all. According to the same formula four digit numbers (words) are 9,000 in all. From this we can see five digit number are 90,000 in all. This numeral theory can be expressed with a simple formula. The formula is - the aggregate of determinable number with whatever digits should be one 0 (zero) less than the digits, whatever quantity they may have, after the digit 9. One zero after a two lettered word (number) 9. Two zero after 9, in case of three digit number, i.e. the sum total of words received from x numbered digit (letter) should be as many as (x−1) numbered zeros are added after 9. It can be expressed in the form of a common formula.

From the above formula it is seen that in math–language two lettered, three lettered as well as two digit, three digit words/numbers are all countable. But in Bengali such word (number) determination is not all possible. Besides it is further noticed – in matters of word in Bengali language words (numbers) having two, three or more letters are not meaningful. But in Mathematics two, three, four, five or more-digit numbers are, meaningful, and they are different from one another in regard of meaning.

Again, the word বসুমতী (vasumati) means পৃথিবী (earth), i.e. the meaning of four – lettered word and three lettered word is the same. In view of constituting separate word is the same. In view of constituting separate meaningful word, it is not unscientific; it is an utter weakness of Bengali language, but in Math – language any two numbers (words) do not mean the same. Each has its own position and has different meaning that is in the scale of meaning each one is original or radical.

In matters of constituting words in Bengali language the more is the number of letters, the deeper is the number of meaningless words. But in math-language the above weaknesses are off with the constituting words (numbers).

In math-language the letters are 0 to 9. Then comes the number 10. Like the rotation of annual season, there come new stems and roots in the rotation of 10 (ten), which are the words free form endings: such as

10, 20, 30, 40, 50, etc. These are stems as well as roots, such as-suppose 20 is a stem as well as root, i.e. with no ending: 20 = 20 + 0.

Then come 21, 22, 23, 24, 25, 26, 27, 28, 29. Here 20 (is a nominal base) with which are added 1, 2, 3, 4, 5, 6, 7, 8, 9 endings respectively and we get 21, 22, 23,............ 29. These are all parts of speech, i.e. parts of speech = words + endings. Similarly 21 = 20 + 1, 22 = 20 + 2, 23 = 20 + 3, 24 = 20 + 4 ...., 29 = 20 + 9: i.e. the basic word is 20 (Nominal). With it, when the endings, 1, 2, 3,.........9 are added, there come new parts of speech, which are capable of use in sentence. Each one is original in the scale of meaning.

So, two–lettered (two–digit) word (number) as well as stems are 10, 20, 30, 40, 50, 60, 70, 80, 90 = 9 in all and two lettered parts of speech are 11, 12, 13, 14,.......19; 21, 22, 23, 24,........., 29; 31, 32, 33, 34, 35,..........39, then upto 99.

One lettered (one-digit), two – lettered (two-digit), three lettered (three-digit) numbers (words), or with more lettered (digit) numbers (words) are all nominal's; i.e. 0 to infinite all integral numbers, whose one place value is zero, belong to nominal's. Similarly, the total numbers of three digit stems are 9, as 100, 200, 300, 400, 500, 600, 700, 800, 900 etc. i.e., if the place value of the first figure of a number is zero, then it is called stem or root.

Now let us see the mystery of constituting word (number): There is no number will stay in the first figure. There is no other complex mystery except taking help of vowels; such as: 7 = 7 + 0 or 7 = 7 × 1.

0 and 1 are vowels. In this case word and parts of speech are the same. Again two lettered (digit) number (word) as well as 10 to 99 = 90 in all; The mystery of constituting these number is that when ten place digit is multiplied by 10 and one place number multiplied by one, and added, then the number is got: e.g.; the ten place digit of 23 is 2 and the one place digit is 3. When 2 is multiplied by 10 and 3 is multiplied by 1, then we get 23.

Chapter – 7

PARTICULAR USE OF WORDS AND PARTS OF SPEECH
(শব্দ ও পদের বিশিষ্টার্থে প্রয়োগ)

Words and parts of speech have literally a particular meaning. But for skilled use the same word expresses different meaning beyond its literal meaning:

Such as – an example of the word কথা (katha/i.e word/speech/talk/tale or tell etc) used in different meaning:

i. রামবাবুর 'কথা' অনেক শুনেছি (I've heard the tale of Rambabu very much).

ii. এসব হল কথার 'কথা' (These are words for words sake).

iii. তুমি 'কথা' দিলে আমার চিন্তা থাকেনা (I need not worry if you promise/commit/assure).

iv. আমি কোন রাজনীতির 'কথা'য় থাকিনা (I don't involve myself in any political affairs).

v. কথায় বলে - নাচতে না জানলে উঠান বাঁকা (There is a saying/proverb '–that a bad workman quarrels with his tools'.

In this way many sentences can be presented using the word কথা (katha). In each case the word কথা (katha) means differently.

Now let us see if in math-language the same word can be used to express different meanings; such as the number (word) 20. As 20 pairs of eggs, 20 dozen of bananas. Ram weighs 20 kgs. Payesh (sweet milked rice) made of 20 liters of milk, travel for 20 miles etc. That is: these first figures are expressing the number 20 sometimes to express numerals, sometimes weight, sometimes an area, again sometimes distance, i.e. in math-language too we notice different meanings of the same word in every step!

That Mathematics therefore is really a language gives birth to no ambiguity.

Chapter – 8

CASE – ENDING (বিভক্তি)

In Bengali grammar a nominal or a root is not generally used in a sentence directly. With this nominal word or root one or more than one meaningless letters are added to make them appropriate parts of speech and help the sentence express properly. This letter or combination of letters is called ending.

That is, the letter or the combination of letters being added with nominal word or root, making parts of speech, is called 'ending' – such as এ (e), কে (ke), রে (re), এর (er); ইব (iba), ইলাম (ilam), ইতেছে (itechhe)....... etc.

Ending are of two varieties: (i) (শব্দ বিভক্তি) Nominal ending and (ii) (ক্রিয়া বিভক্তি) Verbal ending. When the endings are added with nominal words, they are called nominal endings: such as রাম + কে = রামকে (Ramke/Ram or to Ram), ঘর + এ = ঘরে (ghare/at home/or in the house), ছাত্র + রা (chhatrara/students)... etc. Again when the endings are added with the roots and make them verbs, they are called verbal endings: such as - দেখ্ + ইব = দেখিব (dekhiba/shall or will see), শুন + ইল = শুনিল (sunila/heard), পড় + ইবে = পড়িবে (padibe/shall or will read)...etc.

In math-language use of similar endings is also noticed. In Mathematical number (word) the letter (digit) which words as ending, is never meaningless. From 0 to 9: 0, 1, 2, 3, 4, 5, 6, 7, 8, 9 – these ten digits are called endings in math-language. 10 or 20, or 30 etc, being added with nominal base as well stems, create different numbers (parts of speech).

The numbers (words) 10, 20, 30,............90, 100,..........., 900 etc, are nominal base, and free from endings: such as 10 = 10 + 0; 20 = 20 + 0; 40 = 40 + 0 these may also be called parts of speech with zero endings. Again, 10 + 1 = 11; 10 + 2 = 12; 10 + 3 = 13; .... 10 + 9 = 19 etc. The above numbers (words) are respectively parts of speech with endings.

E.g. 10 added with 1 ending makes the new parts of speech = 11

10 added with 2 ending makes the new parts of speech = 12

10 added with 3 ending makes the new parts of speech = 13

.... etc.

The number (word) next to 19 is 20, which is a nominal base or nominal, the parts of speech of the 20's are – 21 = 20 + 1; 22 = 20 + 2; 23 = 20 + 3........... 29 = 20 + 9.

Thus ending is added with the nominal 20, and the new part of speech 21 comes of. 2 endings are added with the nominal 20 and the new parts of speech 22 comes of ......etc.

That is the original word (number) is 20, 30, 40,.............. 90, 100, 300, ...900.

With them 1, 2, 3,........9 etc. letters (digits) being added in the one place (or first figure) position, are created as nominal (case) ending or new parts of speech. Innumerable examples may be gives like the above ones.

Again, the simplest value of $(7 \times 3) + 2$ is a new number as well as word. But 7×3 is a common multiplication. With that 2 is added to make a new number (word) = 23.

Similarly $(5 \times 7) + 3$ typed mathematical expressions are the examples of verbal ending. The first figure (one place digit) of any number is really comparable (or equivalent) to ending.

Again, in math-language it is written as $2x + 3y = 70$.

The above equation (in sentence) x and y are nominal's, because variable number may be x or y; as it may be age, weight etc. of a person, and may be the index of amount or, of distance (or difference). So, with the nominal's x and y the endings 2 and 3 are multiplied to befit as parts of speech in a sentence (equation), and to help the sentence-express itself. That is, in a mathematical equation (sentence) the co-efficient of the variable numbers (word or number) are the application of the verbal endings.

In a word, 0 to 9 digits are the endings. That is the math-alphabet (the digits) contains ten digits, as: 0, 1, 2, 3, 4, 5, 6, 7, 8, 9. They are the signs of endings. It is very easy to get by heart. They are not so complex as the letter or letters of Bengali grammar, such as এ (e), কে (ke), রে (re), ইলাম (ilam), ইতেছি (itechhe), ইবে (ibe) etc. That is, in math–language there is no use of endings like those of Bengali language.

Then, what can we say in a word?......... it is – 'Mathematics is a language', or it is capable of expressing our feelings exactly like Bengali or any language. This idea can be extended for all languages.

Chapter – 9

COMPOUNDS (SAMASA) (সমাস)

Compound means mixing or connecting. Compound is created in order to express language in a concise form. When two or more parts or speech having mutual relation or affinity, are unified, then it is called compound. Now let us notice compound in Math- Language.

Avyayibhava Compound (অব্যয়ীভাব সমাস)

In the sense of want (অভাব অর্থে): e.g. ভিক্ষার অভাব (want of alms) = দুর্ভিক্ষ (famine); বিঘ্নের অভাব (want of interruption) = নির্বিঘ্ন (uninterrupted); ভাতের অভাব (want of rice) = হাভাত (utterly wretched person); মিলের অভাব (want of accord) = গরমিল (discord).

In math-language: 19 = কুড়ির অভাব (want of/less than twenty); 29 = তিরিশের অভাব, (want of/less than thirty); 39 = চল্লিশের অভাব (want of/ less than forty); 49 = পঞ্চাশের অভাব (want of/less than fifty) etc. ঊন (Ūna) means অভাব বা কম (want/less). In mathematics if the first figure is 9, then it can be termed as অব্যয়ীভাব সমাস (Indeclinable compound). That is 19, 29, 39, 49,...... etc, numbers (words) suffers from the want or incapability of turning into the original word or base word to 20, 30, 40, 50... etc. Besides in math– language any number (word) can be expressed in the form of অব্যয়ীভাব সমাস (Avyaybhavasamasa). Such as 12 = তেরোর অভাব (want of/less than thirteen); 15 = ষোলোর অভাব (want of/less than sixteen); 24 = পঁচিশের অভাব (want of/less than twenty five).

In the Sense of Unsurpassability (অনতিক্রম অর্থে)

E.g. শক্তিকে অতিক্রম না করে (not surpassing/exceeding power) = যথাশক্তি (as far as power permits); বিধিকে অতিক্রম না করে (not surpassing the law/rule) = যথাবিধি (according to law);

In math-language; $x < 7$ not exceeding 7, $x < 20$: not exceeding 20.

The above type of in-equation may be expressed in the form of Avyayibhava compound.

In the Sense of Till/Until/Upto (পর্যন্ত অর্থে)

In Bengali শৈশব পর্যন্ত (till childhood) = আশৈশব (till childhood); পা থেকে মাথা পর্যন্ত (from foot to head) = আপাদ মস্তক (foot to head); মরণ পর্যন্ত (till death) = আমরণ (till death); সমুদ্র পর্যন্ত (up to the sea) = আসমুদ্র (upto the sea), জানু পর্যন্ত (up to the knee) = আজানু (upto the knee).

In math-language: $1 + 2 + 3 \ldots\ldots + 30$.

Sum from 1 to 30. $1 + 2 + \ldots 30 = \sum_{r=1}^{30} r$

$\sum_{r=1}^{n} r^2 = 1^2 + 2^2 + 3^2 + \ldots\ldots n^2$ ... Summation of square's from 1 to n

$S = \{x : 1 < x < 7\}$, x an integer.

$S = \{2, 3, 4, 5, 6,\}$ i.e. 2 to 6 all are integers.

$S1 = \{x : 10 > x > 3\}$, X are integers. $= \{4, 5, 6, 7, 8, 9\}$

That is, these are the integral numbers which range from 4 to 9. Like the above examples all the sum totals or sets can be expressed in the sense of till/until in the indeclinable compound.

In the Sense of Contrariness (বৈপরীত্য অর্থে)

E.g. কূলের বিপরীত (opposite to the shore/flow) = প্রতিকূল (to the contrary); ফলের বিপরীত (opposite to the result/consequence) = প্রতিফল (reverse action/result), দানের বিপরীত (opposite to gift/giving away)= প্রতিদান (return gift).

In math–language: $x > 0$; $x < 0$ opposite to it, $x < 0$; $x > 0$ opposite to it.

In the Sense of Frequence/Repetition (বীপ্সা-অর্থে)

In Bengali language - রোজ রোজ (everyday) = হররোজ (everyday/daily), দিন-দিন (everyday) = প্রতিদিন (everyday/daily); সন-সন (every year) প্রতিসন (each/every year).

In math-language: $7 + 7 = 2 \times 7$, i.e seven & seven = double seven (in the sense of addition).

Multiplication process is the miniature form of addition; i.e. repeated addition of one type of number can be expressed in the form of Avyayibhava compound. And $7 \times 7 = 7^2$; $3 \times 3 = 3^2$, $5 \times 5 = 5^2$ this type of expression is also included in indeclinable compound.

In the Sense of Similarity (সাদৃশ্য অর্থে)

E.g. দ্বীপের সদৃশ (like an island) = উপদ্বীপ (a peninsula/a mini island); বনের সদৃশ (like a forest) = উপবন (a grove/garden), ভাষার সাদৃশ (like a language) = উপভাষা (a dialect).

In math-language: $\triangle ABC \sim \triangle PQR$

i.e. $\triangle ABC$, $\triangle PQR$ are similar

$\angle ABC$ angle similar to $\angle PQR$ etc. That is in every line of Geometry the use of Avyayibhava compound is seen in the sense of similarity as proof of resemblance.

Tatpurusa Compound (তৎপুরুষ সমাস)

Karma Tatpurusa (কর্ম তৎপুরুষ) compound is defind with the following words – অতীত (Atit/passed over/surpassed), আগত (Agata/come to), সংক্রান্ত (Sankranta/relating), অতিক্রান্ত (Atikranta/exceeded) etc.

In Bengali Language = সংখ্যাকে অতীত (exceeding the number) = সংখ্যাতীত (Sankhyatita); প্রশ্নকে অতীত (over bearing the question) প্রশ্নাতীত = (Prasnatita); কল্পনাকে অতীত (over reaching imagination) = কল্পনাতীত (kalpanatita); স্বপ্নকে অতীত (over reaching dream) = স্বপ্নাতীত (swapnatita).

In Math language: S = {x:x > 7}

That is, the set indicates those numbers which are beyond 7, the numbers are greater than 7. Similarly, all the sets like the above forms are the examples of karma Tatpurusa (কর্ম তৎপুরুষ) Samasa. (i.e Accusative Tatpurusa). Again in Bengali শাসনকে সংক্রান্ত (Relating to law/administration) শাসন সংক্রান্ত; বিষয়কে অতিক্রান্ত (exceeding the subject/topic) = বিষয়াতিক্রান্ত etc.

Again in Mathematics S = {x: x > 5}

That is – those numbers exceed 5.

Likewise all the sets as above are the examples of karma Tatpurusa (কর্ম তৎপুরুষ)

And in Bengali: শরণকে আগত (come in shelter/lodged in shelter) = শরণাগত, মরণকে আগত (succumbed to death) = মরণাগত।

In Mathematics 23 = 24 আগত (ensued); 35 = 36 আগত (ensued). That is any mathematical integral number (word) is less than the ensuring number (word). So, all the numbers can be expressed in similar karma Tatpurusa. And in Bengali in the use of the word, সংক্রান্ত (relating); ধর্মকে সংক্রান্ত (relating to religion) = ধর্ম সংক্রান্ত, ঘটনাকে সংক্রান্ত (relating to the incident) = ঘটনাসংক্রান্ত।

In Mathematics: y = f(x); x সংক্রান্ত অপেক্ষক (relating function)

= f(y); y সংক্রান্ত অপেক্ষক (relating function)

= f(t); t সংক্রান্ত অপেক্ষক (relating function)

All the above function can be expressed as karma Tatpurusa with relation. And in Bengali ছেলেকে ভুলানো (tocojole the boy) = ছেলেভুলানো, গাঁটকে কাটা (cutting the pursue) = গাঁটকাটা (pursue- cutter).

In Mathematics: 2 × = Multiplication of two, 5 ÷ = Division of 5, 3 × = Multiplication of three;

9 ÷...... = Division of 9;

All the examples are of Karma Tatpurusa.

The Tatpurusa in which endings of karan karaka (instrumental case): দ্বারা (by), দিয়া (with), কর্তৃক (with the help of) is called karan Tatpurusa; e.g. তুষার দ্বারা জীর্ণ (worn out bysnow) = তুষারজীর্ণ; ঢেঁকি দ্বারা ছাঁটা (husked by husking pedal) = ঢেঁকিছাঁটা; গোঁজা দ্বারা মিল (solution by improper means) = গোঁজামিল; শ্রী দ্বারা যুক্ত (added with beauty) = শ্রীযুক্ত।

In mathematics: ÷ Y 3 কে (Multiplication of 3) দ্বারা (by) 2 = 3 × 2

÷ Y 5 কে (Multiplication of 5) দ্বারা (by) 3 = 5 × 3

$5 \div 2 = 5 \times \dfrac{1}{2}$ = ÷ Y 5 কে (Multiplication of 5) দ্বারা (by) $\dfrac{1}{2}$

2 + 2 + 2 + 2 = 2 × 4 = ÷ Y 2 কে 4 দ্বারা (Multiplication of 2 by 4.

That is- analysis from Mathematical point of view –any addition, subtraction, multiplication, division or any kind of processing or, all mathematical operational signs are the examples of karan Tatpurusa (instrumental case).

Apadan Tatpurusa (AblativeTatpurusa) (অপাদান তৎপুরুষ সমাস)

The Tatpurusa compound in which ablative endings (i.e.হইতে, থেকে, চেয়ে, অপেক্ষা/from) become extinct, is called অপাদান তৎপুরুষ/ablative Tatpurusa; e.g.পদ হইতে চ্যুত (fallen from position) = পদচ্যুত; বিলাত হইতে ফেরত (returned from England) = বিলাত ফেরত; শাপ হইতে চ্যুত (released from curse) = শাপমুক্ত; আদি হইতে অন্ত (from beginning to end) = আদ্যন্ত।

In Mathematics-

5 – 2 = 5 হইতে 2 এর অন্তর (difference from 5 to 2)

7 – 3 = 7 হইতে 3 এর অন্তর (difference from 7 to 3)

7 – 3 > 0, 3 অপেক্ষা 7 বৃহত্তর (7 is bigger than 3)

x < y; y অপেক্ষা x ক্ষুদ্রতর (x is smaller than y)

Non-Tatpurusa (নঞ্তৎপুরুষ)

The Tatpurusa compound, in which the preceding word is a negative indeclinable, is called নঞ্তৎপুরুষ (Non Tatpurusa) compound, e.g. নয় ঐক্য (no unity) = অনৈক্য (Disunity); নয় আবাদি (not cultivable) = অনাবাদি (uncultivable); নয় ইষ্ট (Not beneficial) = অনিষ্ট (harmful); নয় আহূত (not called) = অনাহূত (uncalled).

In Mathematics-

নয় ধনাত্মক (Not positive) = ঋণাত্মক (Negative) i.e. $x > 0$ or $x < 0$

নয় ঋণাত্মক (Not Negative) = ধনাত্মক (positive) i.e. $x < 0$ or $x > 0$

নয় মূলদ (Not Rational) = অমূলদ (Irrational)

x নয় মূলদ মানেই, (x not rational means) = x অমূলদ (x Irrational).

(Karmadharaya) Compound (কর্মধারয় সমাস)

মধ্যপদলোপী কর্মধারয় সমাস (Karmadharaya compound with extinct/elided middle word):

The Karmadharaya compound in which middle word of the expounding sentence becomes extinct/elided from the compounded words, is called Madhyapadalopi karmadharaya; e.g. সিংহ চিহ্নিত আসন

(Lion marked chair) = সিংহাসন (throne/royal chair); ঘি মিশ্রিত ভাত (ghee – mixed rice) = ঘি-ভাত; জল মিশ্রিত দুধ (water mixed milk) = জলদুধ; ঘরে পালিত জামাই (fostered son in law) = ঘরজামাই।

In Mathematics 23 = Two ten three (দুই দশ তিন)

32 = three ten two (তিন দশ দুই)

47 = four ten seven (চার দশ সাত)

37 = three ten seven (তিন দশ সাত)

Except base word as well as 10 20 30 40 etc, numbers (words;) all the numbers can be expressed in the form of middle word extinct/elided Madhyapadalopi karmadharaya (মধ্যপদলোপী কর্মধারয়) compound.

Dwandwa Compound (দ্বন্দ্ব সমাস)

The compound in which both the preceding and the succeeding words retain their meanings after being compounded, is called Dwandwa compound,

e.g. ভাই ও বোন (brother and sister) = ভাইবোন; জায়া ও পতি (wife and husband) = দম্পতি (couple); কাজ ও কর্ম (work and task) = কাজকর্ম; মাতা ও পিতা (Mother and father) = মাতাপিতা (parents).

In Mathematics-

25 = 20 & 5 (joining as well as in the sense of addition).

35 = 30 & 5 (joining well as in the sense of addition)

27 = 15 & 12 (joining as well as in the sense of addition)

That is, in mathematical problem any kind of addition can be expressed in the form of Dwanda (দ্বন্দ্ব) compound.

Bahubrihi Compound (বহুব্রীহি সমাস)

সহার্থক বহুব্রীহি (Bahubrihi compound in the sense of accompaniment):

The Bahubrihi compound, in which the accompanying nominal word proceeds, is called সহার্থক বহুব্রীহি. (Bahubrihi compound with an accompaniment compound); i.e. স্ত্রীর সহিত বর্তমান (accompanying wife) = সস্ত্রীক (with wife); অর্থের সহিত বর্তমান (accompanying money/means) = সার্থক (successful)

শ্রদ্ধার সহিত (with respect)= সশ্রদ্ধ (respectful); সম্মানের সহিত বর্তমান (with honour) = সসম্মান (honourable).

In Mathematics-

32=30 with 2 (30 এর সহিত 2 বর্তমান) that is 32 = 30 + 2; 45= 40 with 5 (40 এর সহিত 5 বর্তমান); 45 = 40 + 5 i.e. any pure mathematical number (word) can be expressed in the form of Bahubrihi (বহুব্রীহি) compound.

Negative Bahubrihi (নঞর্থক বহুব্রীহি)

The Bahubrihi compound in which the preceding word is a negative indeclinable (অব্যয়) is called Negative Bahubrihi Compound; e.g. অর্থ নাই যার (That which has no meaning or money) = অনর্থ (Meaning less); শোক নাই যার (The person who has no grief/mourning) = অশোক (griefless); নাই খোঁজ যার (the person who has no trace) = নিখোঁজ (trace less); নাই লজ্জা যার (the person who has no shame) = নির্লজ্জ (shameless).

In Mathematics-

39 = 1 less than 40 (40 থেকে 1 কম যার)

49 = 1 less than 50 (50 থেকে 1 কম যার)

29 = 1 less than 30 (30 থেকে 1 কম যার)

25 = 1 less than 26 (26 থেকে 1 কম যার)

In math-language all the numbers (words) can be expressed in the form of Negative Bahubrihi (নঞর্থক বহুব্রীহি) compound.

VyatiharBahubrihi Compound (ব্যতিহার বহুব্রীহি সমাস)

The Bahubrihi compound in which actions are mutual or exchangeable, or in which the same word being doubly used, constitutes both the preceding and the succeeding words, is called Vyatihar Bahubrihi; e.g. লাঠিতে লাঠিতে যে যুদ্ধ (a fight with sticks) = লাঠালাঠি; হাতে হাতে যে যুদ্ধ (a fight with hands) = হাতাহাতি; কানে কানে যে কথা (whisper heard with ears) = কানাকানি; গলায় গলায় যে প্রেম (love hugging each other's neck) = গলাগলি; দলে দলে যে বিরোধ (feud between parties/groups) = দলাদলি।

In Mathematics-

$3 + 3 = 2 \times 3 = 6; 2 + 2 = 2 \times 2 = 4; 7 + 7 = 2 \times 7 = 14$

All the above examples, i.e. three and three – two into three (six); two and two = four; seven and seven, two into seven (fourteen) – are similar examples to those of Vyatihar Bahubrihi compound; that is – the similar numbers being added twice or more, give the result (aggregate) by way of addition, and are worked out in the same method as that of Vyatihar Bahubrihi compound.

Chapter – 10

JOINING (সন্ধি)

Joining means mixing or coinciding. Rapid pronunciation in order to concise language, joining of two or more sounds for their proximity, elision or change are called sandhi (Joining): e.g. দেব + আলয় = দেবালয় (temple); জীব + অণু = জীবাণু (germ/bacteria). Rule for the joining (স্বরসন্ধি/ swarasandhi) of words, the preceding of which ends with a vowels, and the succeeding of which begin with a vowel: অ + অ = আ, আ + আ = আ, ই + ই = ঈ etc.

Rule for joining the words, the preceding of which ends with a consonant and the succeeding of which begins with a consonant: ক + ই = কি, ক + ঈ = কী, ক + ঐ = কৈ etc. Besides there are নিপাতন সিদ্ধ (irregular) joining, বিসর্গ সন্ধি joining (i.e. joining of words, the preceding of which ends with বিসর্গ (ঃ); So, in Bengali grammar joining are bound in a large number of rules. Naturally, valuable time is, therefore, misused to keep the rules in memory.

But alongside this, in Math-language use of joining can be clearly seen in times of mathematical calculation: e.g. 2 + 5 = 7 or 3 + 2 = 5 – these are the examples of letter joining; i.e. with the joining of two letters a new meaningful letter is created. It is also the example of letter joining (বর্ণসন্ধি). Again 15 + 12 = 27, 17 + 10 = 27………. etc. With the joining of two words a new word (number) is created. It is an example of word joining. Again 5 × 7 = 35 or 35 ÷ 7 = 5 = these are also included in joining. As multiplication is the miniature form of addition, so the product of multiplication is called rapid joining, and 35 ÷ 7 = 5, it is called a very rapid subtraction-joining.

There is weakness in Bengali regarding addition-joining: e.g. সিংহ + আসন = সিংহাসন (throne/lion-marked chair). In math-language the world of joining is free from this weakness: e.g. $21 + 11 = 32$. Again $11 + 21 = 32$, That is commutative property of addition holds in mathematics but not in Bengali. Also more, associated property hold in addition in mathematical language, but not in Bengali: e.g. $(12 + 24) + 11 = 12 + (24 + 11) = 47$.

But in Bengali language, such use of words is not expected anywhere in joining. It is a great weakness in constituting a language, while in math-language for joining all the children acquire the general rules of addition, subtraction, multiplication and division as well as joining in their childhood. Time is not wasted as there are no long and separate rules.

So, in Bengali language as well as in math-language joining is not painful, rather scientific. In math-language, therefore the use of joining of Bengali grammar is also noticed; it is not burdened with rules.

Chapter – 11

SENTENCE, SUBJECT & PREDICATE OF A SENTENCE ARRANGING PARTS OF SPEECH IN A SENTENCES
(বাক্য, বাক্যে উদ্দেশ্য-বিধেয় ও বাক্যে পদস্থাপন)

When some meaningful words added with ending and sitting side by side as parts of speech and express our feelings, and also compel the listeners and readers to react, then it is called sentence: e.g. 'Now Madhu is passing through good time.' Or 'Your handwriting is not good.' These are complete sentences; i.e. these sentences are able to express mental feeling completely.

Besides, there are incomplete sentences – e.g. the age of Ram and Shyam differs. It is an incomplete sentence, because the sentence does not express the age – difference of Ram and Shyam; i.e. when the sentence ends, the listeners natural curiosity is not satisfied; and he remains curious to know the answer or listen to the second part of the sentence. In math-language also the use of complete and incomplete sentences (equation & in-equation) are seen.

E.g. $x-2y$ is an incomplete sentence. It has been expressed by stating the difference of twice the age of son (y) from the father's age (x). But what is the difference? – The answer to this question remains unknown. Similar numbers (incomplete sentence, polynomial) all are incomplete sentences i.e. in math-language there is use of incomplete sentence (polynomial/in-equations). Again the number $\{(4 \times 3) - 5\}$ is an incomplete sentence. And the number (x–3) is also an incomplete sentence as well as polynomial.

Again, $y = 2x + 1$ is an equation as well as a sentence. Grammatically it is a complete sentence as it is capable of expressing mental feeling fully. With this equation multipurpose relations can be expressed: That is, a complete sentence is an equation in math-language. Besides in our conversation we often express another kind of sentence: e.g. 'The money power of the Tatas is mightier than that of the Birlas; or Ram eats more rice than Shyam; or my age is more than yours; These statements are written is Mathematics as: $x > y$, where x, y are the money-power of the Tatas and the Birlas; or the amount of rice eaten by Ram and Shyam; or the age indicator of you and me. Thus all the equations can be expressed in language. The statements are also contrarily true.

That is, complete sentence incomplete sentence and unequal sentence (in-equation) of the world of language, are called in math-language, equations polynomial and in-equations respectively. Again in some sentences are noticed- 'Birds fly in the sky (পাখি আকাশে ওড়ে) – in this sentence, the word আকাশ (sky) being added with এ (e-endings, i.e. আকাশে) has become a part of speech to be used in sentence.

Similarly, in math-language, in a sentence (equation) as well as $2x + 3y = 60$, in this equation x & y are pronominal words, which are the definite expression of whether it is of age of a person or of distance, x & y are used in lieu of them. Simply, x & y are pronouns, the sum of twice of x & thrice of y are equals to 60. Being added with verbal endings 2 & 3 are parts of speech and fill for use in sentence (equation). That is, in math-language any equation is a complete sentence. So in comparison with Bengali language Mathematics is also a complete language.

Subject and Predicate of a Sentence

A sentence has two parts: Subject and Predicate. The person or thing about whom or which something is said, is called subject, and what is said about the subject is called predicates, e.g. The boy who had come, went away – Here the boy is Subject, and the rest part (i.e. had come …….. went away etc) is Predicate.

In math-language too a sentence is made with the combination of subject and predicate: e.g. the aggregate of the age of Ram and Shyam is equal to 30, or $x + y = 30$. The sum total of the age of Ram and Shyam as well as $(x + y)$ is subject and equal to 30 is predicate.

In a math-language (equation) there are two parts – one is L.H.S (left hand side) and the other is RHS (right hand side). The left hand side is the subject and the right hand side is the predicate; that is, $A - 2B = 5$; in equation $(A - 2B)$ is subject and 5 is predicate.

Thus all the equations can be divided into two parts, of which one is subject and the other is predicate.

Arrangement of Parts of Speech in Sentence

General Rules for arranging Parts of Speech in Sentence:

1. Subject is placed in the beginning of a sentence and the finite verb at the end.
2. In case of transitive verb the subject precedes the verb.
3. Infinite verb precedes the finite verb.
4. There should be relation between subject and verb.
5. For the clearness of long sentence there should be punctuation marks etc.

Let us see if similar rules can be applied in mathematical sentence. For the benefit of making it clear let us takes an equation:

Suppose: $[(2x - y) \div 2]$ of $1/3 = 20$

In this case $(2x - y)$, the number is the subject and comes in the beginning of the sentence. Division process ($\div$) as well as finite verb is placed after the subject $[(2x - y) \div 2]$. The infinite verb is set in the place of the following verb before এর (i.e. of). The sentence is complete, and as it assures a relation, so there is consistency between the subject and the verb. Again, the subject being placed near the verb, has maintained the explicitly of meaning, and assured respite/pause of sentence by maintaining punctuation marks in proper place and brackets. That is, to make the meaning of a sentence clear by using punctuation marks is also noticed in Mathematical equation.

In this way proof can be placed by giving a large number of examples to fulfill similar conditions. So, in a word it can be said 'Mathematics is a scientific language.'

Chapter – 12

VOICE CHANGE (বাচ্য পরিবর্তন)

Before going to change voice, it is necessary to know what voice is. Voice is formed in combination of subject, object and verb. The role of the three parts of speech in framing sentences is most important.

In a sentence the relation of verb with that of the subject, object and other parts of speech, or relative mode of speaking is called voice: e.g. kuber catches hilsa fish or hilsa fish is caught by kuber, or kuber's catching of hilsa fish.

The above relation of verb with subject, object and other parts of speech, or relative mode of speaking is called voice.

The framing of sentence changes as a result of the change in mode of speaking. The change from one voice to another voice is called 'voice change'.

Some rules must be followed in changing voice: as

a. If the active voice is in chaste language, the passive voice will also be in chaste language.

b. In the subject of the active voice are added the postpositions: কর্তৃক, দ্বারা, দিয়া (by/with by adhering র, এর (i.e. s/of)

c. The object of active voice is added with the ending (অ) of the passive voice.

d. In case of verbs with double objects, the direct object (i.e. মুখ্যকর্ম) generally becomes the subject. The indirect object (গৌণকর্ম) is retained (i.e. retained object/অনুক্তকর্ম)

e. Some often the subject is added with ending: for example:

Active voice   – Valmiki composed the Ramayana.

Passive voice   – The Ramayana was composed by Valmiki.

Active voice   – You have destroyed the Mewar.

Passive voice   – The Mewar has been destroyed by you.

Active voice   – Ramchandra killed Ravana.

Passive voice   – Ravana was killed by Ramchandra.

Let us watch voice change in matter of construction and language vis-a-vis the above theories and examples. In Math language any sentence as well as equation is taken, such as

$y = 2x + 1$ …… (1)

This is a definite relation between two numbers. The above equation (sentence) can be expressed in changed sentence differently, where the original sense remains the same: as –

$y = 2x + 1$ …… (1)

$2x - y + 1 = 0$ …… (2)

$x = \dfrac{1}{2}(y - 1)$ …… (3)

$2x = y - 1$ …… (4)

The equations (sentences) (1), (2), (3), (4) are the same equations (sentences), but expressed differently. In-equation no. (1), the relation of doubling the verb and adding 1 of subject (y) with the object (x); i.e. makes a nice complete sentence (equation) of relation between y & x. As the value of y is related by x, it is called the function of x, and expressed y = f(x). This can be termed as active voice at par with that (i.e. active voice) in Bengali grammar.

Again equation no. (3) is an expression of different type of direct relation of object x and subject y. In this case the value of the variable number x is half of the number (y – 1). Here also x is called the function of y, and expressed by x = f(y). That is, if each function is taken for active voice, then its opposite function should be passive voice: e.g.:

Active voice: $y = 2x + 3, y = f(x)$

Passive voice: $x = \frac{1}{2}(y - 3), x = f(y)$

Active voice: $5x - 3y + 3 = 0$ or $y = \frac{1}{3}(5x + 3) = f(x)$

Passive voice: $3y - 5x - 3 = 0$ or $x = \frac{3}{5}(y - 1) = f(y)$

Active voice: $\frac{x}{2} - \frac{y}{3} = 1$ or $y = \frac{3}{2}(x - 2)$ or $y = f(x)$

Passive voice: $x = \frac{2}{3}(3 - y)$ or $x = f(y)$

Active voice: $y = \sin x$ or $y = f(x)$

Passive voice: $x = \sin^{-1} y$ or $x = f(y)$

Active voice: $y = \cos x$ or $y = f(x)$

Passive voice: $x = \cos^{-1} y$ or $x = f(y)$

Active voice: $y = \log x$ or $y = f(x)$

Passive voice: $x = e^y$ or $x = f(y)$

From the above examples it is noticed that the rules of voice-change followed in Bengali grammar, are applied similarly in the mathematical voice–change too. In this matter the general rules of math-language addition (+), subtraction (−), Multiplication (×), division (÷) or the transposes, square, square root, cube, cube root……. etc are applied.

It can be summarily admitted that the mathematical equation s fulfill a step further in establishing mathematics as a language with the support of voice change.

Chapter – 13

TENSE (কাল)

Let us notice the three sentences:

 i. Ram played. (রাম খেলা করেছিল)

 ii. Ram is playing. (রাম খেলছে)

 iii. Ram will play. (রাম খেলা করবে)

The three sentences above are the expression of three different times. In the first sentence the verb 'played' makes it clear that the work of playing was accomplished earlier. In the second sentence the verb 'is playing' means that the work of playing is continuing at present. And in the third sentence the verb 'will play' shows that the work of playing has not yet begun. It will begin in future. That is, the verb play occurred, is occurring or will occur in different times. So, the time, when the action happens, is called the time of verb or 'Tense'. According to the difference of time the finite verb is divided into three: **(a) Past Tense (b) Present Tense (c) Future Tense**.

Alongside this, if we watch keenly, we'll see that in math-language too the impact of tense persists. If it is said: three years ago the age of father was double the age of his son. It is an incident of past. The present age of father is y and the present age of son is x. It is expressed in math-language: $y - 3 = 2(x - 3)$. Here the expression $(y - 3)$ is the father's age three years ago. The expression $(x - 3)$ is the son's age three years ago. After simplification of the above equation it becomes $y = 2x - 6 + 3 = 2x - 3$. That is, the past incident can be said with present reference-

'the present age of father is 3 years less than the double of son's age.' That is, in math-language any sentence (equation) of the past is expressible in support of the present, which is never possible in Bengali or English or in any grammar.

Now let us refer to an incident: such as the present speed of the bus is five times than of the tram. The present speed of the bus is y, the present speed of the tram is x. Then it can be expressed in math-language as $y = 5x$.

Now let us refer to an incident of future, such as- 'after five years Ram's weight will be three times than Shyam's weight'. In this case Ram's present weight is y-unit and Shyam's present weight is x-unit. Then in math-language it can be written- $y + 5 = 3(x + 5)$. The expression $(y + 5)$ is the mathematical expression of Ram's weight which will be after five (5) years. The number $(x + 5)$ is the mathematical expression of Shyam's weight which will be after five years.

The simplified form of the above equation is; $y = 3x + 15 - 5 = 3x + 10$. That is the future incident can be said in support of the present- Ram's present weight is 10 kg. more than three times weight of Shyam. That is, in math-language any equation (sentence) of future tense is expressible in support of present, which in a word, is not possible in Bengali or English or in any grammar.

That is in math-language – clear stepping in present, past and future. It is further noticed that Math-language is not confined to the compass of time. Occurrence of each tense in support of the present can be determined nicely and definitely.

Chapter – 14

PUNCTUATION MARKS IN A SENTENCES (বাক্যে অভ্যন্তরীন যতিচিহ্ন)

At the time of speaking, the speaker has to express his/her feeling taking care of two aspects- (i) breathing and (ii) expressing feeing, so that his/her spoken language is justified in respect of meaning and depth, and easily expressed. When we read or speak a sentence, we need some pause/stop after reading or speaking the complete sentence. This pause/stop is called punctuation or punctuation mark.

For the sake of pause/stop within the sentence we use interim pause/stop, e.g. He said, "I have no pen". Or "Coming back he saw that he had gone." The punctuation marks are generally comma (,), semi colon (;), dash (–), full stop (.), inverted comas ('…') etc.

That is the interim punctuation marks are the rest within the sentence, which give perfection to a sentence, and help the sentence express the right meaning. Otherwise the meaning of the sentence changes completely. The sentence ends with a full stop, which gives the sentence perfection. Full stop is the ultimate declaration of the end of the sentence.

Now let us see if there is any use of punctuation in math-sentence. To bring out the amount of rupees it is written: 1, 32, 52, 327. The sum of rupees is 1 crore, 32 lakh, 52 thousand, 3 hundred, 27 (twenty seven). To utter the sum, the punctuation- comma and full stop have been used in the places where the stroke of breathing is required.

That is, in order to express a big number punctuation marks are used after the places where the reading of a place value ends (,) and at the end full stop (.) is used. In this case as well as to express a very big number a common formula is used. At first a comma should be used after three digits from right to left, and then the commas should be used gradually towards left after every two digits, whatever big the size of the number is, the same rule should be followed; i.e. in math-language to read and understand correctly a big number regarding mathematical punctuation-marks are properly used.

Besides, in some mathematical problems- e.g. the thrice of the difference of age of Ram and Shyam is equal to 20. In the problem, if the present age of Ram and Shyam is x & y respectively, then the difference of their age is $(x - y)$. At the time of reading till the difference of age between Ram and Shyam there is a definite relation, again in its multiplicity of thrice, there is a special relation and at the end equals to twenty. As a punctuation mark is necessary for a particular pause, so also in math-language, in order to differentiate the relation through brackets, punctuation is used: e.g. $(x - y) \times 3 = 20$. As pause is often given through punctuation mark to read and understand a language nicely, so also in math-language in expressing mathematical equation (sentence) punctuation marks like comma or various types of brackets, full stop etc. are often used for understanding, writing or solving any problem.

Chapter – 15

SUBSTITUTION OF SINGLE WORD OR, ONE-WORD EXPRESSION
(এক কথায় প্রকাশ)

We express our mental feeling through sentence. Collection of words to compose that sentence, grammatical correctness and laconic presentation make the sentence an art. Shortening of sentence began to value modern life. Condensation of sentence came into being in order to beautify a sentence. This shortening of sentence is otherwise called- 'integration of sentence' or 'one-word expression' or 'substitution of single word': That is, unification of many parts of speech is named as sentence- condensation; or 'signification of parts of speech' or 'one-word expression'. For example- the person who knows grammar-'grammarian'; existing in the same time-'contemporary'; the person whose lineage and character are unknown–'stranger'; the son-in-law nurtured/domesticated at his father-in-law's house-'domestic son-in-law', …. etc. The need of shortening is inborn in man. One-word expression ensures from that need.

In math-language too the same use is seen. To the ordinary learner's mathematical language is equivalent to horror. In order to exempt from horror the least, one-word expression emerged in lieu of enlarged sentence. Not only that, one-word expression has come off to offer contraction of time, beauty of expression and collective manifestation of artistic attitude or manner.

As for example, to express the ultimate value of a expression, we frequently use |x| which means both the numbers- either positive or negative, and again, perhaps – neither.

That is; $|x| = x$; when $x > 0$

$ = -x$; when $x < 0$

$ = 0$; when $x = 0$

∴ To mean a number positive or negative or neither, x is expressed in a word.

Again, the sum total 1 to 100 is naturally expressed $= \sum_{r=1}^{100} r$

One word expression of a few such mathematical subjects are given below:

$1 + 2 + 3 + 4 + 5 + 6 \ldots + (n-2) + (n-1) + n = \sum_{r=1}^{n} r$

$1^2 + 2^2 + 3^2 + 4^2 + 5^2 + 6^2 \ldots + (n-2)^2 + (n-1)^2 + n^2 = \sum_{r=1}^{n} r^2$

$1^3 + 2^3 + 3^3 + 4^3 + 5^3 + 6^3 \ldots + (n-2)^3 + (n-1)^3 + n^3 = \sum_{r=1}^{n} r^3$

$a^2 + 2ab + b^2 = (a + b)^2$

$(a+b)(a-b) = a^2 - b^2$

$a^3 + 3a^2b + 3ab^2 + b^3 = (a + b)^3$

$a^n + n_{C_1} a^{n-1} b^1 + n_{C_2} a^{n-2} b^2 + \ldots n_{C_r} a^{n-r} b^r + \ldots b^n = (a+b)^n$

$\sin \alpha + \sin(\alpha + \beta) + \sin(\alpha + 2\beta) + \sin(\alpha + 3\beta) \ldots + \sin\{\alpha + (n-1)\beta\} = [\sin[\alpha + (n-1)]]$

$\cos \alpha + \cos(\alpha + \beta) + \cos(\alpha + 2\beta) + \ldots \cos\{(\alpha + (n-1)\beta\} = [\cos[\alpha + (n-1)]]$

$$a + (a+d) + (a+2d) + (a+3d) \ldots + \{a + (n-1) = [2a + (n-1)d] =$$

a → fisrt term,

l → last term, n → number of terms.

$$a + ar + ar^2 + \ldots + ar^{n-1} = \text{ when } r>1$$

$$= \text{ when } r<1$$

Sin A Cos B ± Cos A Sin B = Sin (A ± B)

Cos A Cos B ± Sin A Sin B = Cos (A ∓ B)

$$\frac{\tan A \pm \tan B}{1 \mp \tan A \tan B} = \tan(A \pm B)$$

$$\frac{\tan A + \tan B + \tan C - \tan A \tan B \tan C}{1 - \tan A \tan B - \tan B \tan C - \tan C \tan A} = \tan(A + B + C)$$

That is, all the mathematical formulas are mathematical one-word expression. Moreover, 2 + 2 + 2 + 2 + … + up to twenty = 2 × 20

Or [{ (20 – 2) – 2} – 2] =[{18 – 2} – 2] = [16 – 2] = 14 This type of all shortening are one-word expression.

That is, the special characteristics of substitution of single word in Bengali are also noticed in mathematical language. So mathematics must be given recognition as a language.

Chapter – 16

INDECLINABLES (অব্যয়)

The general meaning of Avyayas (অব্যয়/indeclinable) is a part of speech which does not decay or change. The definition of indeclinable is – the words used in sentences do not change their gender, number, or endings, i.e. the indeclinable are not subject to change of their original forms.

Suppose, a mathematical equation is taken $2x + 3y = 70$. Here it is a complete sentence as well as an equation. The synthesis of both the parts of speech is made by the sign (+), Again $4 + 4 = 8$; $14 + 12 = 26$.

In the first of the two above examples a new letter (digit) is created with the assimilation of two letters; a new part of speech (number) has emerged as a result of assimilation of the second two words (parts of speech). As the addition signs maintain mutual consistency, so they are considered indeclinable.

Mathematical signs, i.e. $+, -, \times, \div$, ... all are indeclinable.

Chapter – 17

ADJECTIVES, ADVERBS AND THEIR CLASSIFICATIONS
(বিশেষণ পদ ও প্রকার ভেদ)

Definition: An adjective is a word that qualifies a noun or pronoun by expressing the vice, virtues, quantity/amount, number, condition etc. of the latter.

And an adverb is a word that qualifies the vice, virtues, quantity/amount, number, condition etc. of adjective, another adverb, verb etc. other than noun or pronoun.

An indeclinable is a word that never changes its form wherever (in a sentence) it is used.

So qualifying words are two in number:

(i) Adjective and (ii) Adverb

(i) Adjective: The word that expresses the vice, virtue, colour, quantity/amount, degree, size etc. is called adjective: such as – The 'beautiful' girls play. The blue clitoria flower is really 'beautiful'.

Here 'beautiful' and 'blue' are adjectives. In math-language there is nice use of adjectives: such as $2x + 3y = 50$; it is a mathematical equation as well as a sentence. With this equation innumerable relations can be expressed such as- The sum total of twice of age of Ram and thrice of age of Shyam is fifty. Here the age representing Ram and Shyam is x & y respectively. These are certainly the expressions of nouns and pronouns. When their age is multiplied twice or thrice, then it is specially qualified, and deserves to be called a clear example of adjective. There are many similar examples can be cited.

Adjectives are of three types:

i. Adjective to Noun

ii. Adjective to Pronoun and

iii. Indeclinable Adjective.

Adjective to Noun or Pronoun

The word which precedes a noun and expresses the vice, virtue etc. of the noun and is called an adjective to the noun: such as- the sky is bright with the luminous stars, i.e. the stars are qualified by the word, 'luminous' so the word 'luminous' is an adjective word to the stars; i.e. an Adjective to Noun.

Likewise in math-language it is said- 'Three less than sixty is equal to double the age of Ram; the mathematical expression of which is: $2x + 3 = 60$. In this case 'x' represents Ram's age. As this representation is a pronominal expression, so in math-language nominal and pronominal adjectives are almost the same. This 'x' either noun or pronoun is specially qualified doubly. It is a numerical adjective; i.e. in any mathematical equation (sentence) the co-efficient of variable numbers are all adjectives to nouns or pronouns.

In pronominal adjective it is noticed- 'the more are the views, the more are the ways; or 'the more you eat, the heavier you become.' These types of expressions are seen in mathematics too: After 2 years Yadu's age will be what Ram's age was 5 years ago. Mathematically it can be expressed as; $x - 5 = y + 2$. As x, y is the present age of Ram and Yadu respectively.

Again, the sale-price of two mangoes is what the buying price of three mangoes is. Mathematically it can be expressed: $3x = 2y$. As x, y are the buying and selling price of a mango, i.e. in math-language no difference is noticed in nominal and pronominal adjectives. So, in view of mathematics monomial expression is more scientific than polynomial expression.

Indeclinable Adjective

In mathematics indeclinable adjective cannot be identified/marked separately: such as – it can be written – 'When Ram went to buy a pencil, he saw- provided he had paid 2 rupees more, he could have bought a pen with the same price'. Mathematically it can be expressed; $x + 2 = y$: there x and y are the price of a pencil or some birds were sitting on a tree: suddenly 5 of them flew away. Then how many were still there? If the number of birds sitting on a tree is x, then $x - 5 = ?$ (How). Thus innumerable examples can be given. But if the equation is thrown earlier, then it is known as indeclinable adjective. But it is difficult to segregate indeclinable adjective from $x - 5=?$

Adverb to Adjective

The word that modifies (or qualifies) an adjective by expressing the latter's vice, virtue, quantity/amount, condition etc. is called an adverb: e.g. Don't overreach. Today is chilly cold. The stars are twinkling in deed dense darkness.

When in math-language it is written in an equation; $x = 2(3y + 5z)$. Here if Ram's weight $= y$, and Shyam's weight $= z$ and Madhu's weight $= x$, it becomes a unit, then it is seen that the weight of x is double of the sum of thrice of y, and five times of z's weight.

The part of speech $(3y + 5z)$ being itself qualified by 3 and 5 respectively, and it is further qualify after multiplying by 2. So it is naturally an adverb to the adjective.

Adverb

The word that qualifies/modifies a verb by expressing the latter's vice, virtue, condition, quantity/amount etc. is called adverb: such as – Walk along the street 'steadily' The train runs 'speedily'. These steadily, speedily – all are adverbs.

In math-language it is written- (2×3); 2, then the adverb is noticed (2×3) is a verb. Multiplication too is a verb. When it is multiplied by 2, then it is qualified, and becomes an adverb; i.e. various uses of adverbs are noticed in mathematical equation or number or in similar expression. If the subject matters, which are dealt in every nook and corner of Bengali grammar, or English grammar, are mentioned at random in math-language too, then mathematics must be a language.

Chapter – 18

AMPLIFICATION (OF IDEAS)
(ভাবসম্প্রসারন)

Poets or story-tellers uphold many elaborate stories and the depth of the stories in a few words artistically and metrically. To express this in lucid language analytically, is called amplification of ideas.

Brief but deep-rooted ideas are contained laconically in a time or two of a poem or prose-piece. To manifest the basic language of the song sung through symbols is called amplification of ideas.

Such as – Hatred burns like straw the person, who commits injustice and bears with it. (অন্যায় যে করে আর অন্যায় যে সহে/তব ঘৃণা যেন তারে তৃণসম দহে।)

And – In this world the person who has plenty, longs for more,

The kind steals the wealth of the poor. (এ জগতে হায়, সেই বেশী চায়, আছে যার ভুরিভুরি/রাজার হস্ত করে সমস্ত কাঙালের ধন চুরি।)

Again – The man who loves man, serves god. (জীবে প্রেম করে যেইজন/সেইজন সেবিছে ঈশ্বর।)

There are many more examples. The above lines are manifestation of deep ideas in rhythmic form. When the basic ideas are amplified, the ordinary people comprehend the idea easily and enjoy.

Now let us see if there is the use of amplification of ideas in mathematics.

Interpretation of: $\int_a^b f(x)dx$. It is a brief and artistic expression of a long matter of discussion. If the matter is expanded, then it is noticed: let f(x) be a single valued, bounded and continuous function defined in the closed interval $a \leq x \leq b$

Now, if we sub divide the interval $a \leq x \leq b$ into n sub interval of equal length of 'h' as

a, a + h, a + 2h, ... a + (n – 1)h; where a + nh = b => nh = b–a

then;

$$\lim_{h \to 0} = h[f(a) + f(a+h) + f(a+2h)....f\{a+(n-1)h\}]$$

$$= \lim_{h \to 0} h \sum_{r=0}^{n-1} f(a+rh); \text{ (a one-word expression)}$$

It is called an integral of the function of $f(x)$ with respect to x within the limit of a & b, and it is expressed by: $\int_a^b f(x)\,dx$ i.e. the above explanation comes off if the symbol $\int_a^b f(x)\,dx$ is amplified. In mathematical language many long speeches can be expressed in a symbolic form under which is hidden long explanation and review.

A few similar examples are given;

1. $\frac{dy}{dx}$'s significance and definition.
2. $\sin(A \pm B) = \sin A \cos B \pm \cos A \sin B$
3. $\cos(A \pm B) = \cos A \cos B \mp \sin A \sin B$
4. Geometric proof of these are clearly amplification of ideas-
5. $(\cos q + i \sin q)^n = \cos nq + i \sin nq$
6. $(a+b)^n = a^n +{}^n c_1 a^{n-1}b^1 +{}^n c_2 a^{n-2}b_2 + ...{}^n c_r a^{n-r}b^r + ...b^n$

7. $(a+b)^2$'s Geometric explanation:

As for example; $(a+b)^2 = a^2 + 2ab + b^2$ - this identity is set up in the student's mind only in the form of an algebraic formula. But it has a geometric figure where it is clearly noticed it is the sum total of two squares and two rectangles. Next the question is – which ones are square and which ones are rectangular? and whether the sum total of the two square areas and that of the two rectangular areas are bigger or smaller? What is the easy sign to identify the square and rectangular areas? In the condition what is the measurement of the areas of the smallest square? These are naturally the question of measurement – that are discussed isolating as another branch of mathematics named mensuration.

As an answer to these questions let us draw a sketch:

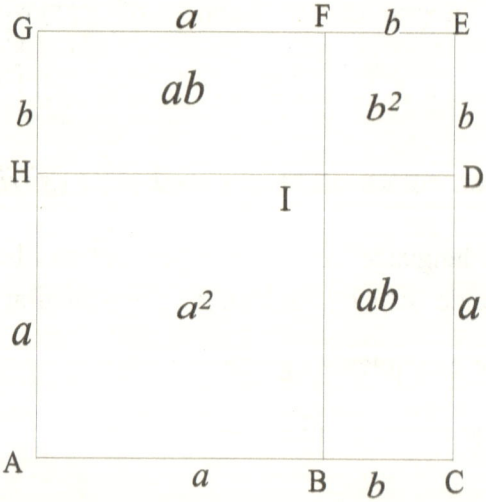

The two squares are a^2 and b^2, the sketch of which are ABIH and DIFE respectively. The two rectangles are $2\,ab$, the sketch of which are BCDI and IHGF respectively, a numeral dimension. Two squares, such as a^2; b^2; two rectangles of two numeral dimensions, such as (ab). And the sum total of two square areas ($a^2 + b^2$) square unit and the sum total of two rectangles 2(ab) square unit and of course the sum total of two area measurements of the square area are bigger than the two area – measurements of the rectangle.

Because, $(a-b)^2 > 0$; a, b real.

$$\Rightarrow a^2 + b^2 - 2ab > 0 \Rightarrow a^2 + b^2 > 2ab$$

And this formula cab be expressed in the form of a building-plan of a low-income group family. a^2 sq. unit is the bed room b^2 sq. unit is the latrine or bathroom. (ab)sq. unit is an open verandah on the south. Another (ab) sq. Unit is the kitchen. Thousands of ideas are hidden in a mathematical formula.

Where presence of variable is one but degree is 2 (two) then it is a perfect square. And when the variables are two and degree is 2 (two) it will be a rectangular area.

7. (1) Geometric explanation of a quadratic equation:

Let us throw our attention to the matter of solving this quadratic equation, $x^2 + 2x = 15$ in geometric view point.

If the sum total of a rectangle having a square area of x-unit length, and a rectangular area of x-unit length, and 2-unit width, be equals to 15. Then what will be the geometric explanation for the solution?

Such as; $x^2 + 2x = 15$

$$\Rightarrow x^2 + 2x + 1 = 15 + 1 \Rightarrow (x+1)^2 = 4^2$$

Now let us draw the figure

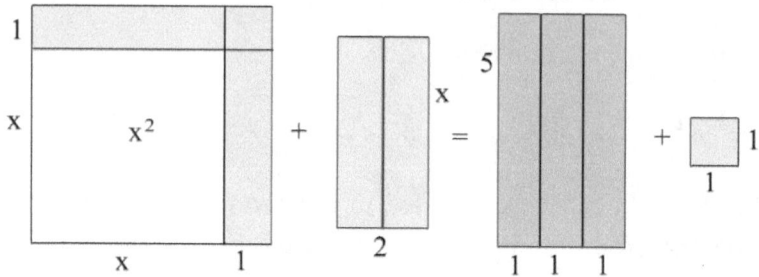

The left hand side of the equation $x^2 + 2x = 15$ is $x^2 + 2x$.

Which is the sum total of a perfect square of length x unit and a rectangle of length x unit and breadth 2 unit. To make left hand side a perfect square we add one square unit on both sides then it becomes:

$(x+1)^2 = 4^2$
$\Rightarrow x+1 = \pm 4$
$\Rightarrow x = \pm 4 - 1$
$= -5, 3$

(ii) Another sketched explanation of a quadratic equation: $x^2 - 6x = 16$

$$x^2 - 6x = 16$$
$$\Rightarrow x^2 - 6x + 9 = 16 + 9$$
$$\Rightarrow (x-3)^2 = 25 = (\pm 5)^2$$
$$\Rightarrow x - 3 = \pm 5$$
$$\Rightarrow x = 8, -2$$

That is, in the depth of each formula there are particular deep ideas hidden in Mathematics to be learnt.

EPILOGUE

"Grammar of Mathematics & Lingual Affinity"- is the first utterance on earth and as an attempt to prove mathematics as a language using its own alphabet, vowel and consonants. From the view point of grammatical alphabet, attempts have been made to make mathematics much easier by presenting mathematics as a clone of a language so that the fear of mathematics can be eradicated from the minds of the math-learner.

REFERENCES

Published in a concise form-

   i. In 'GANIT PRARIKRAMA' – a mathematical journal published by the joint patronage of University of Dhaka & Jahangirnagar University with financial assistance of Bangladesh Govt.

   ii. 9th State Science & Technology Congress – Viswa Bharati University West Bengal.

   iii. 'GYAN-BIGNAN' – the first science journal in Bengal by Satyendra Nath Bose.

   iv. 'Simanta Bangla' - local newspaper.

   v. As a book in Bengali – 'GANITER BYAKARAN O BHASAGATA SADRISHYA'

Copy Right

Mousumi Mitra (wife), Mouli & Mohona (daughters), Rajesh Bhattacharjee & Subhamay Sen (sons in law), Maharaj.

www.ingramcontent.com/pod-product-compliance
Lightning Source LLC
Chambersburg PA
CBHW021005180526
45163CB00005B/1898